More Power to You

Other books by Vicki Cobb

Chemically Active! Experiments You Can Do at Home
Making Sense of Money
The Monsters Who Died: A Mystery about Dinosaurs
Science Experiments You Can Eat
Magic . . . Naturally: Science Entertainments and Amusements
More Science Experiments You Can Eat
Bet You Can! Science Possibilities to Fool You (with Kathy Darling)
Bet You Can't! Science Possibilities to Fool You (with Kathy Darling)
The Secret Life of Cosmetics
The Secret Life of School Supplies
How to Really Fool Yourself: Illusions for All Your Senses
The Secret Life of Hardware: A Science Experiment Book
Lots of Rot
Fuzz Does It!
Gobs of Goo
How the Doctor Knows You're Fine
Brave in the Attempt: The Special Olympics Experience
Supersuits
The Scoop on Ice Cream
Sneakers Meet Your Feet
The Trip of a Drip

More Power to You

by Vicki Cobb
illustrated by Bill Ogden

Little, Brown and Company
Boston   Toronto   London

Text copyright © 1986 by Vicki Cobb
Illustrations copyright © 1986 by Bill Ogden
All rights reserved. No part of this book may be reproduced in any form or by any electronic or mechanical means including information storage and retrieval systems without permission in writing from the publisher, except by a reviewer who may quote brief passages in a review.

Library of Congress Cataloging-in-Publication Data
Cobb, Vicki.
  More power to you!

  (How the world works series)
  Summary: Explains electric power and other forms of power, answering such questions as "How does electric power make a light turn on?" Includes experiments and tricks.
  1. Electric power — Juvenile literature. [1. Electric power. 2. Power resources] I. Title. II. Series.
  TK148.C53 1986   621.31   85-23926
  ISBN 0-316-14899-7

10 9 8 7 6 5 4 3

BP

Published simultaneously in Canada
by Little, Brown & Company (Canada) Limited

Printed in the United States of America

This series is dedicated to Louis Sarlin,
   the teacher who gave me the best year of my childhood
   and the key to my place in the world.

Contents

1. Power at Your Fingertips _____ 1
2. Two Big Ideas about Energy _____ 6
3. How to Make Electricity _____ 12
4. Fossil Fuels _____ 20
5. Nuclear Energy _____ 29
6. Money and Power _____ 39
   How to See a Magnetic Field _____ 47

The author gratefully acknowledges the interest and assistance of Barbara Luce of the Connecticut Yankee Energy Information Center. She would also like to thank Jacqueline Harris and Dale Sigler of Northeast Utilities and Janet Silver of Edison Electric Institute for their help. The author assumes full responsibility for the accuracy and conclusions expressed in the book.

More Power to You

1. Power at Your Fingertips

You have power at your fingertips! You can make a dark room bright with light. You can fill a silent room with music. You can make a blank TV screen come alive with moving pictures. All it takes is the flick of your finger. Electricity waits behind each switch. When you turn it on, you change your world.

Ever turn on a switch and see nothing happen? You might try the switch a few times. You might figure it's the bulb, or the TV's broken, or a fuse blew. It's hard to believe that electric power isn't there. Electric power is something everyone takes for granted.

There are times, though, when power failures occur. An entire neighborhood can be without electricity. Every once in a great while there is a blackout in a big city. These blackouts make the headlines. No one has lights — indoor lights, traffic lights, streetlights go out. Refrigerators and freezers turn off. Food melts and spoils. Electric stoves don't work. During a blackout you can't watch TV or use a washing machine or dryer. You can forget about using the electric can opener and pencil sharpener and hair dryer. The only things you *can* use are run by batteries, or have their own power source, like the telephone, or don't use electricity. A blackout usually shuts just about everything down.

Schools and stores close. The world comes to a standstill.

The world did not always depend on electricity. The first electric lights were lit only about one hundred years ago. Before that people used candles or gas or oil flames for light. They burned wood or coal in their stoves. Clothes were washed by hand. Sewing machines were run with a foot pedal. People read books or visited each other instead of watching TV. There were no speeding

cars, so traffic lights weren't needed. Most food was prepared fresh. Ice cream and other frozen foods were almost unheard-of. It wasn't until 1882 that the first power plant to make electricity was built. Now we can't live very well without electricity.

No one *wants* to. So, we all have to pay for it. Electricity costs money. Every month your family gets a bill from your local power company. The amount of the bill depends on how much electricity was used that month. If the bill isn't paid, the power company can turn off your electricity.

The price of electricity is going up. When you grow up, you will pay more than your parents pay now. And the power at your fingertips costs in more ways than one. Perhaps you've heard of the "energy crisis" and "pollution." The future of your electric bill, the energy crisis, and pollution are all connected. I'll tell you more about this later.

The world is now set up to put electric power at your fingertips. But there are problems that must be solved to make sure you will always have electricity. You can understand these problems.

How will our great need for electricity change the future? How does the world work to bring you electricity? How can you help keep power at your fingertips?

To find out the answers to these questions and more, read on.

2. Two Big Ideas about Energy

A rolling stone, a ray of light, a bowl of hot soup, and a song in the air are all alike in a special way. Surprised? What can such different things have in common? They all have *energy*, that's what. Energy comes in different forms.

Anything that moves has energy of motion. Scientists call energy of motion *kinetic* energy. Wind and waterfalls have kinetic energy. There is kinetic energy in traffic, in horse races, and in a rolling stone. When you move, you have kinetic energy.

Light is another kind of energy. There is light that you see, but there is also some light, like X rays, that you can't see. All light is energy.

Heat is also energy. Hot things, like a bowl of soup or boiling water, have more heat energy than colder things. Heat energy moves from hot things to cooler things that touch it. A lot of hot things lose heat to the air. That's why hot soup cools when you blow on it.

Sound is energy. Sound you can hear is air that moves in an organized way. If a sound is loud enough, it can make things shake. It can even break your eardrum.

And, of course, electricity is energy. But we're getting ahead of our story.

Now here's the first big idea about energy. Are you ready? Here goes. ONE KIND OF ENERGY CAN

BE CHANGED INTO ANOTHER KIND OF ENERGY. Prove it yourself. Rub your hands together quickly. Feel them get warm. You are changing kinetic energy into heat energy. Whenever two things rub together, heat is one result. You can also do the opposite and change heat into motion. Simply boil water. As the water heats up, it starts moving. It moves quite fast when it boils.

A flame gives off both heat and light energy. In fuel, there is stored energy called *chemical* energy. Chemical energy is released by adding a little bit of heat, just enough to get a fire started. Then the heat released by the fire keeps the fire going until the fuel is used up. There is chemical energy in the food you eat. Your body releases this energy as heat and as kinetic energy. It also uses this energy to help you grow and stay healthy.

Here are some things that change one kind of energy into another.

All of these inventions use energy to do some-

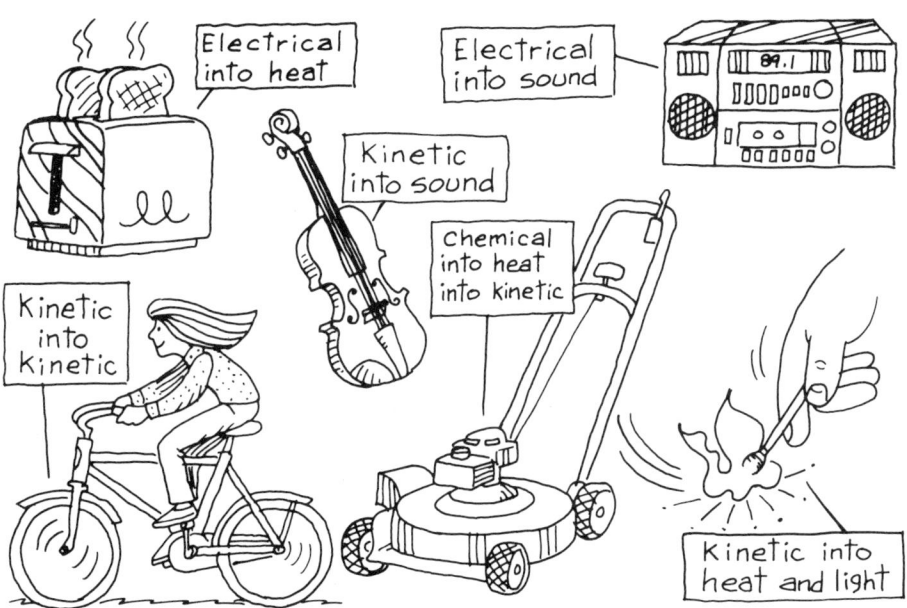

thing people want done. This brings us to the second big idea about energy: ENERGY CAN BE MADE TO DO USEFUL THINGS. There's a lot of energy out there in nature. Winds blow. Rivers flow. The tides come in. The sun shines. If only we could catch this energy and make it do what we want! Easier said than done.

For thousands of years sailors have used the kinetic energy of the wind to move sailboats where they want to go. Steam engines turn the chemical

energy of a fuel into the kinetic energy of a locomotive. But nowadays we have found that the easiest kind of energy to use for doing useful things is electricity.

There isn't much electricity running loose in nature. Lightning is about it. But lightning is almost impossible to control. We couldn't count on it to run anything even if we could catch it. But we have figured out how to make electricity by changing other kinds of energy in nature.

Electricity is energy that is produced in power plants and delivered directly to your home faster than you can blink your eyes. The next chapter tells you how it's made.

3. How to Make Electricity

You can make electricity very easily. Here are some ways you can try. Rub a balloon with a sweater. The rubbed balloon will stick to the wall. Rub a piece of newspaper with a plastic bag. The newspaper will stick to the wall. What makes the balloon and newspaper stick to the wall? You guessed it. Electricity.

When you rub certain objects with certain materials, you give them a special power. They can now attract other things they are not touching. Rub the balloon again. Pass the rubbed balloon over the hairs on your arm, without touching your skin. You can feel the electricity as the hairs of your arm are attracted to the balloon. By rubbing the balloon you have given it an electric *charge.* A charged object will attract certain uncharged objects that are nearby. The area of attraction around a charged object is called an *electric field.* You felt the electric field around the charged balloon with your arm hairs. This electric field is a very weak field. A strong electric field can give you a nasty shock.

Lots of things can get an electric charge. If you have long hair, you can give it a charge by combing it with a rubber or plastic comb. You can give yourself an electric charge by rubbing leather soles against a rug. Point your charged finger toward a metal lamp base or radiator. A small spark will jump from your finger to the metal. Sometimes clothes from the dryer stick together with electricity. This is more likely to happen if you dry man-made fabrics and don't use a fabric softener. It is also much easier to make electric charges when the weather is cold and dry. If the weather is humid, the charges quickly disappear into the air.

This kind of electricity stays in the charged object until it disappears into the air or the ground or some other object. It is called *static electricity*. The word *static* means "standing still." Static electricity can't be used to run anything.

If you bring some metal near static electricity, the electricity jumps to the metal. But the metal itself does not become charged. Instead, the electricity quickly moves through the metal. Metals are called *conductors* because they "conduct" the electricity from one place to another. Electricity moves from one place to another through wires, which are long, thin pieces of metal. Moving electricity is called *current electricity*. Current electricity is the energy you buy from a power company. An electric current moves through wires so quickly that there is no wait at all for it to come to your house from a power plant. It doesn't matter how far away the plant is, either. Throw the switch, and electricity is there. Current electricity waits behind the switches and outlets in your house.

Power companies use a very special metal to make current electricity. This metal is a kind of

iron. You couldn't tell it was a special kind of iron just by looking at it. There is only one way to detect its powers: by bringing it near other pieces of iron and steel. This special iron attracts the other pieces of metal and they wind up sticking to it. You can feel the attraction when you pull them off. This special kind of iron is none other than a *magnet.* The area around the magnet that pulls on metals is called its *magnetic field.*

How can a magnet make electric current? That's the big idea behind every power plant in the world. A *MOVING* MAGNETIC FIELD MAKES AN ELECTRIC CURRENT IN A NEARBY WIRE. If you pass a magnet across a wire, you create a tiny bit of electric current. Keep the magnet moving back and forth and you keep making electric current. A stronger current can be produced by moving the magnet across a lot of wire. That's why a coil of wire is used. A coil is many turns of wire. The more turns in a coil, the stronger the current. Power plants make a lot of electric current by using a large coil made of thousands of turns of wire and very strong mag-

A Simple Generator

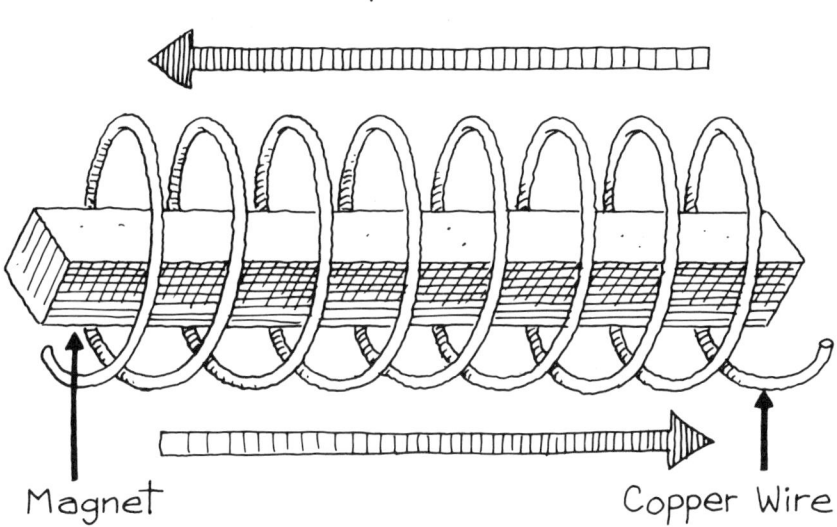

Magnet   Copper Wire

nets. The faster the magnets turn inside the coil, the more electricity is made. The arrangement of moving magnets and wire coil is called a *generator*. A generator changes the kinetic energy of a moving magnetic field into electricity. In this way a power plant generates enough electricity to turn on a city of light bulbs.

Now the problem is how to make the magnet move and keep moving. The solution is as simple as a pinwheel. A pinwheel has blades like a fan that are shaped to catch the wind. It turns when wind pushes against its blades. In a power plant, the magnets in the generator are on one end of a long, strong shaft. There are many blades on the other end of the shaft. These blades are shaped to move when they are hit by moving steam or moving water. The arrangement of shaft and blades is called a *turbine*. The moving steam or water pushes the blades of the turbine. The shaft turns. When the shaft turns, the magnets turn, and electricity is produced.

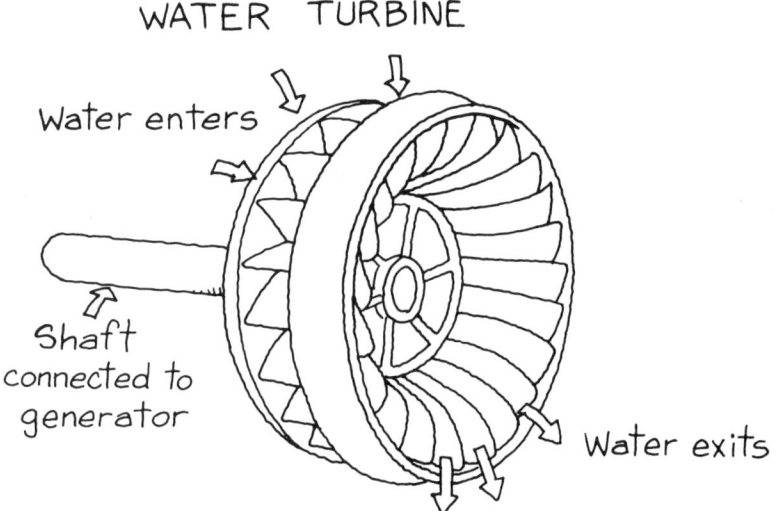

There are some places in the world where the flow of water is strong and steady enough to generate electricity. Niagara Falls is one such place. The power of the falling water turns giant water turbines. A power plant that uses flowing water as its energy source is called a *hydroelectric* plant. *Hydro* means "water." Only a small part of our electricity is made in hydroelectric plants. Most of our electricity comes from *steam-electric* plants. Our energy crisis and pollution problems are a result of the ways power plants make steam.

Keep reading. There's more to this story.

4.
Fossil Fuels

Wind can turn a generator. The blades of a windmill are like a turbine. Windmills can make enough electricity for a few houses. But wind is usually not strong and steady. Even the most modern windmills can't catch enough wind to light up cities. So most power plants make their own strong, steady wind. It's called *steam*.

All you have to do to make steam is boil water. Power plants do this in their huge *boilers*. But boiling water in a power plant is not like boiling water in a pot at home. Instead of letting the steam escape, boilers keep it trapped for a while. The pressure of the steam pushing against the boiler walls grows as more and more steam is made. When the pressure is high enough, a valve opens. High-pressure steam rushes through a pipe into the steam turbine. The strong, hot wind pushes

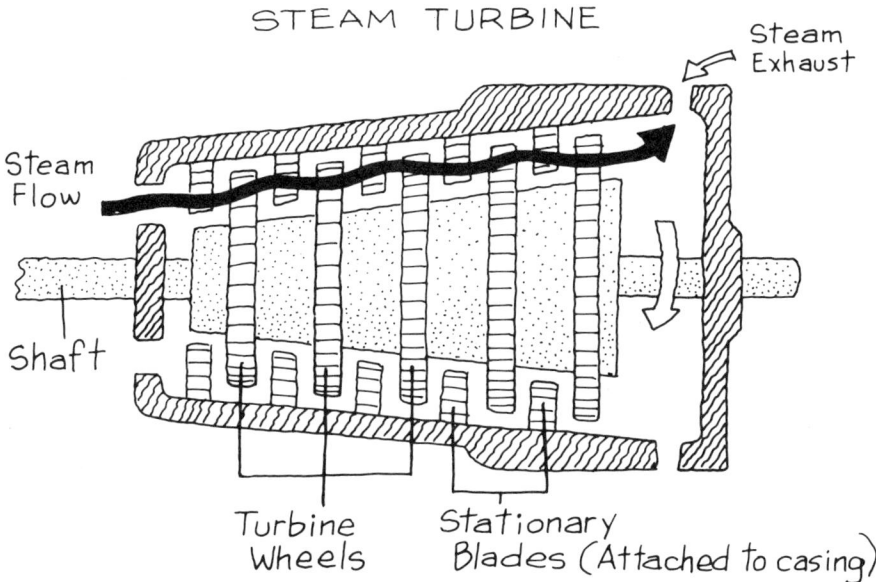

the blades of the turbine. The shaft turns. The generator turns. Electricity is produced.

Power plants make electricity seven days a week, twenty-four hours a day. They need fuel to keep the boiler fires burning — lots of it. Trainloads of coal, tankers of oil, and pipelines of natural gas are used, day after day, at power plants. Electricity costs money because it costs money to build power plants. It costs money to keep them in good condition. But mostly electricity costs money because fuel costs money. And each kind of fuel costs us in other ways as well.

Coal is burned in about half of all power plants. Coal is a soft black rock that is dug out of the earth. Coal is made up of the remains of plants that lived millions of years ago. Most of the plants that eventually became coal grew in swamps. As the plants died, they piled up on the ground. Over the years they rotted and became a thick layer. The layers of dead plants became buried under sand. The weight of the sand pressed down on them. First they turned into a brown, crumbly material called *peat.* Millions of years later, peat became coal. Fuels that have formed in the earth from living things are called *fossil fuels.*

Some coal lies close to the surface of the earth. Heavy machinery is used to strip away the thin layer of rock that covers the coal. A coal-digging machine then scoops up the exposed coal and loads it into a truck. This kind of coal mining is called *strip mining.* Strip mining ruins the land. It leaves behind ugly wounds in the earth. A few years ago a new law was passed to make sure

strip-mining companies repair the damage. Land that has been strip-mined now must be turned into farmland or parks after the coal has been removed.

Coal is also dug out of underground mines. Underground mining is dangerous. Tunnels can collapse. Coal in the mines can catch on fire. Sometimes there are poisonous gases near coal. Modern mining methods have made it less dangerous. Machines now do all the digging and loading. But every once in a while there is an accident that makes news.

When coal burns, it produces gases and smoke. They pollute the air. Some of this air pollution comes back to earth as *acid rain.* One shower of acid rain does not hurt anything, but years of acid rainfall can kill fish in lakes. It can eat away leaves and ruin the soil for farming. Some coal-burning power plants have "scrubbers" in their smokestacks. Although scrubbers trap some of the air pollutants and cut down on acid rainfall, now there is the problem of getting rid of the stuff in the

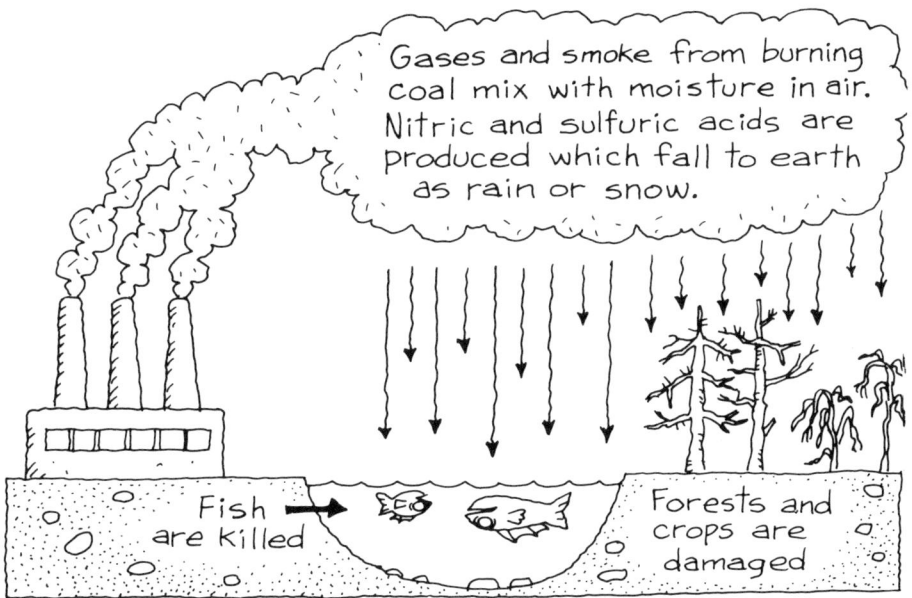

scrubbers. If it is buried, it will pollute the land. Burning coal is dirty, no two ways about it. But we have plenty of coal, enough to keep the power-plant fires burning for at least 1,000 years. The problem is how to stop it from polluting.

Natural gas, on the other hand, is an almost perfect fossil fuel. It is found trapped in rock formations near oil. It can be delivered to power plants easily through pipes. Much less natural gas than coal is needed to make a hot boiler fire. And natural gas burns cleanly. It does not add to the acid-rain problem.

Then why not use natural gas? There isn't much of it around. Experts say we have enough for only about twenty years. Only a few power plants burn it. Natural gas cannot be a big part of energy-making plans for the future.

Oil is a third fossil fuel. Like coal, it formed from once-living things. Oil is different in that it is formed from dead animals and plants that piled up on the floors of ancient seas. The dead sea animals were buried under sand and rock. Pressure over millions of years changed the dead sea creatures into oil and natural gas. These are trapped underground in rocks or in underground pockets.

Oil is produced by drilling wells through the earth. Pumps bring oil to the surface. Pipes and oil tankers deliver oil to power plants.

Oil is a clean fuel. Burning oil does not pollute

the air. But some people claim that the world supply of oil will last about twenty years. Much of the oil used in the United States comes from other countries. It is brought here on ships called oil tankers. Such imported oil is expensive. Sometimes there is an accident and a tanker spills its oil. Spilled oil can hurt sea life and ruin beaches.

The energy crisis is about running out of oil and natural gas. People are busy searching for new

supplies. Burning coal adds to the problems of air and water pollution. People are busy thinking of safer, cleaner, and cheaper ways to supply energy to generate electricity.

For a while it looked as if atomic energy was the answer. Maybe it is one. That's what the next chapter is about.

5. Nuclear Energy

There are mines in the southwest United States where giant earth-moving machines remove a crushed gray rock. The rock contains traces of a metal called *uranium*. The crushed rock must be put through several steps to remove the uranium and turn it into a special kind of fuel. Uranium becomes fuel for a different kind of power plant.

Uranium does not burn like fossil fuels. The energy in uranium is stored inside the very smallest piece of uranium, the uranium *atom.*

No one has ever seen an atom. Atoms are much too small. But scientists have many other ways of learning about them. We know that there are ninety-two different kinds of atoms on earth. Each kind of atom makes up a different substance called an *element.* Uranium is the natural element with the heaviest atoms. Hydrogen is the element with the lightest. Oxygen is a light element. Gold and lead are heavy. In their laboratories, scientists have made elements that are heavier than uranium, but these elements do not exist naturally on earth.

All atoms are alike in certain ways. Atoms are not solid balls. Most of the weight of an atom is at its center, called the *nucleus* (NOO-clee-us). Very tiny, lightweight particles travel at high speeds around the nucleus. These particles are called *electrons.* Electrons are like planets in orbit. The nucleus is like the sun. There is a lot of space between an atomic nucleus and its electrons.

The nucleus of an atom is made of two kinds of particles, *protons* and *neutrons*. Hydrogen, the simplest atom, has a nucleus of only one proton. One kind of uranium has a nucleus of 92 protons and 143 neutrons. When you add 92 and 143, you get 235. So this kind of uranium is called U^{235}. U^{235} is used as fuel in power plants. Another kind

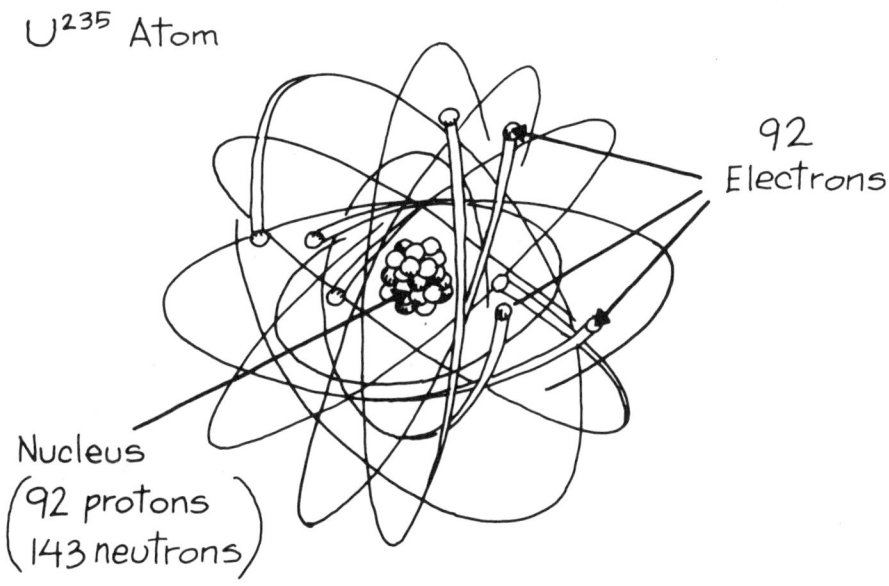

U^{235} Atom

92 Electrons

Nucleus
(92 protons
143 neutrons)

of uranium has 92 protons and 146 neutrons in its nucleus. It's called U^{238} and it is not useful as fuel. Most of the uranium that is mined is U^{238}. Only a very tiny amount is U^{235}.

The protons and neutrons in an atomic nucleus are held together with energy called *nuclear energy*. It takes so much energy to hold the heavy nucleus of the uranium together that eventually the uranium nucleus crumbles, or "decays." As it breaks down, it gives off small amounts of nuclear energy in the form of radiation that is like X rays. This kind of energy is called *radioactivity*.

You cannot detect the radioactivity without special instruments. A tiny amount of radioactivity, called *background radiation,* is always present and is quite harmless. But large amounts of radioactivity can make people sick. How sick you get depends on how much radioactivity you are exposed to and how long you are exposed. Huge amounts in a short period of time can cause death. Exposure to small amounts of radioactivity over a long period of time can cause cancer. So around radioactive substances there is lots of safety equipment and there are many rules for safety. People who work with nuclear fuel and people who work in nuclear power plants are very well protected against radioactivity. In one year they are exposed to the same amount of radioactivity you might get if you had a few chest X rays.

Radioactive uranium does not give off enough nuclear energy to run a power plant. So scientists have figured out another way to make uranium

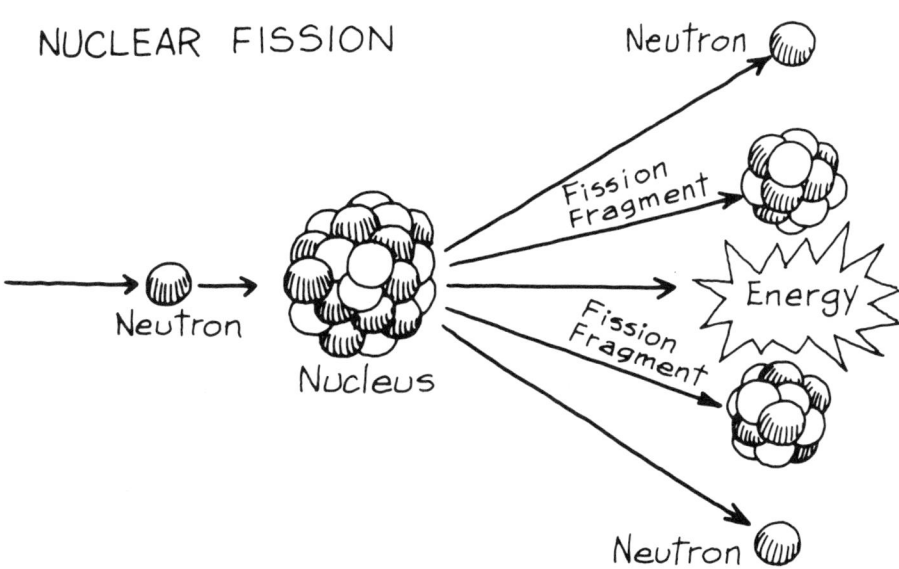

NUCLEAR FISSION

release lots of nuclear energy. Here's how. Every once in a while, a decaying uranium nucleus gives off a neutron. This extra neutron is like a bullet. If it strikes another uranium nucleus, it can split the nucleus apart. Splitting the heavy uranium nucleus is called *nuclear fission*. When nuclear fission takes place, an enormous amount of heat is released, along with lots of radioactivity. The split nucleus also releases two or three extra neutrons. These extra neutrons can strike other *nuclei* (NOO-clee-eye, meaning more than one nucleus) and split them. This is a *chain reaction,* in which splitting nuclei cause more nuclei to split, and on and on.

When nuclear fission occurs in an atom bomb, all of the heat and radioactivity is released in a split second. But in a power plant, nuclear fission is controlled. The uranium fuel is in tiny pellets about the size of the tip of your little finger. The pellets are in rods about twelve feet long. About 200 rods are packed together in a bundle. Bundles are put in the *core* of the reactor. Water is pumped

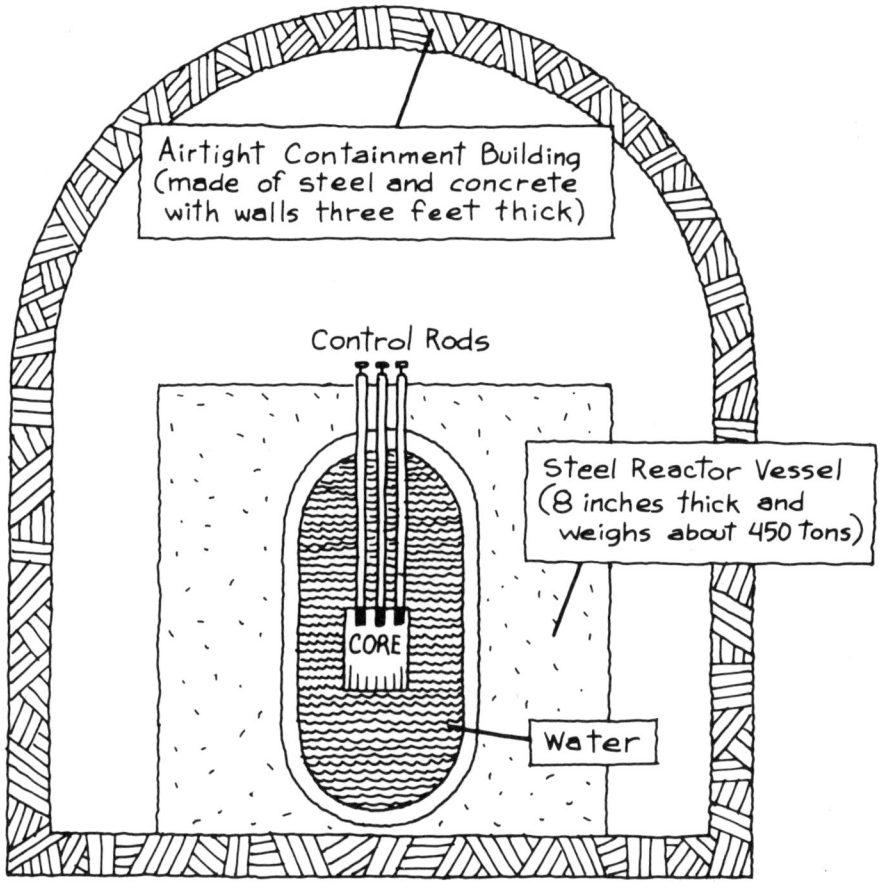

around the core. The heat of nuclear fission heats the fuel rods, which in turn heat the water. It becomes steam, and the steam turns the turbines of the generator.

The water also contains a substance that catches extra stray neutrons. This removes many of the "bullets," so that the reaction doesn't get out of control and get too hot. There are also lots of control rods sitting next to the core. If there is an emergency, the control rods are pushed into the core, where they absorb the neutrons. This stops the nuclear reaction.

A bundle of nuclear fuel rods lasts about three years. When it is removed from the core, it is very hot and more radioactive than before. It is stored underwater in special pools. After about eight months it is safe enough to be taken out and packed in cases that will not let the radiation escape. The fuel pellets can go to a special factory where the remaining useful fuel can be removed and made into new pellets. But there is radioactive waste left over. One problem that has to be solved is what to do with the radioactive waste. It must be disposed of safely so that radiation does not leak into the air, water, or ground.

At one time it looked as if nuclear energy would

be the main source of energy for the future. Nuclear energy is clean. No pollution comes out of a nuclear power plant. One ton of uranium fuel gives off as much energy as 14,000 tons of coal. And uranium fuel is fairly inexpensive. In addition, there's enough around to last at least 500 years. Scientists have also developed ways to make useful fuel from U^{238}, and the used pellets can be recycled.

But it now looks as if nuclear energy is not the answer to our hopes for future energy. A nuclear power plant is very expensive to build. Many safety features are needed to make sure radiation does not escape into the air. Every step during construction has to be approved by the government. This takes extra time. As time passes, costs go up. Some nuclear power plants have been under construction for fourteen years when they could have taken only six, and they turned out to be much more expensive than anyone expected.

What's the biggest problem with nuclear energy? Most people do not understand it. They are afraid of it. People think of the atom bomb that killed thousands. They do not realize that it is *impossible* for a nuclear power plant to explode like a bomb. They do not understand that nuclear energy is at least as safe as energy from fossil fuels. It just has to be properly handled. And we do know how to handle it.

Many countries in the world are sold on nuclear energy and are ordering new plants all the time. But in the United States politicians and the public spend a lot of time arguing over nuclear energy. Nuclear plants are becoming too expensive to build.

One thing is clear. We can't make electricity in the future the way we're making it now. Now is the time to think and to plan.

6.
Money and Power

You agree to pay for electricity every time you turn on a switch. Somewhere in your house or apartment building there are meters that measure how much electricity you use. Your electric company sells you electricity by the *kilowatt-hour*. You use one kilowatt-hour to burn ten 100-watt light bulbs for one hour or to run your color TV for four hours. The meters in your home keep track of how many kilowatt-hours your family uses. Every month or so someone from the electric company comes to read your meters.

The amount paid for one kilowatt-hour depends on where you live. Electricity is more expensive in some places than in others. But if you don't like the price, you can't shop around for cheaper electricity. When you shop for sneakers or ice cream, you can compare prices from different companies. If you wish, you can buy the cheapest brand. But there is only one company where you live that sells power. It has no competition to keep its price down.

What keeps the electric company from raising its prices sky-high, if it wants to? The government, that's what. The government names a group of people to be an *agency* for the public. The agency acts like a watchdog. It controls the price of the kilowatt-hour. Electricity is too important to everyone and too expensive to make to let just anyone produce it and charge what they want.

Electric companies serve all people. As a result, people have the right to know how they run their business. If an electric company wants to build a new plant, they go to the energy agency in their area. The agency holds *public hearings.* At public hearings some speakers tell why the plant should be built. Others tell why it should not be built. Anyone can go to these hearings and make up his or her own mind. Sooner or later people decide by voting.

People may be unhappy about their electric bills. They may argue about building a new power plant or a nuclear waste plant. Not everyone will be happy no matter what is decided. But there are three things that everyone agrees about. First, we can't go backward and live without electricity. Second, someday we will run out of fossil fuels. Third, solving the problem of energy in the future is going to cost money.

All of the electric companies in this country give money to a fund that pays for research. Researchers experiment with new ways to use other kinds of energy to turn generators. Some of the ideas are already being put to work in power plants. Others are still on the drawing board. The people who are thinking about our future energy needs are *engineers*. It is the job of an engineer to solve a problem to make something work. We will need lots of good engineers in the future. Perhaps you would like to be one. Here are some of their ideas.

If you dig deep enough into the earth, it gets very hot. Heat beneath the earth's surface is called

geothermal ("earth's heat") energy. In a few places on earth, geothermal energy is very close to the surface. One such place is in northern California. The rocks are so hot there that they produce steam naturally. This steam runs turbines. No fuel is needed in their plants! But geothermal energy near the earth's surface is hard to find. And we haven't yet figured out how to use the deep heat we really have to dig for.

What about the sun? *Solar* energy is free wherever the sun shines. How can solar energy be made into electricity? One way is to collect sunlight with mirrors. The mirrors shine a beam of sunlight at a "power tower." The heat produced by all the reflected sunbeams makes steam at the power tower. Another way to use solar energy is to make giant solar batteries. A battery is stored electricity. Solar batteries are used in space to provide energy for satellites and spaceships.

There are problems with solar energy. The mirrors and power-tower idea needs lots of sunlight and lots of land. It will not work well when the

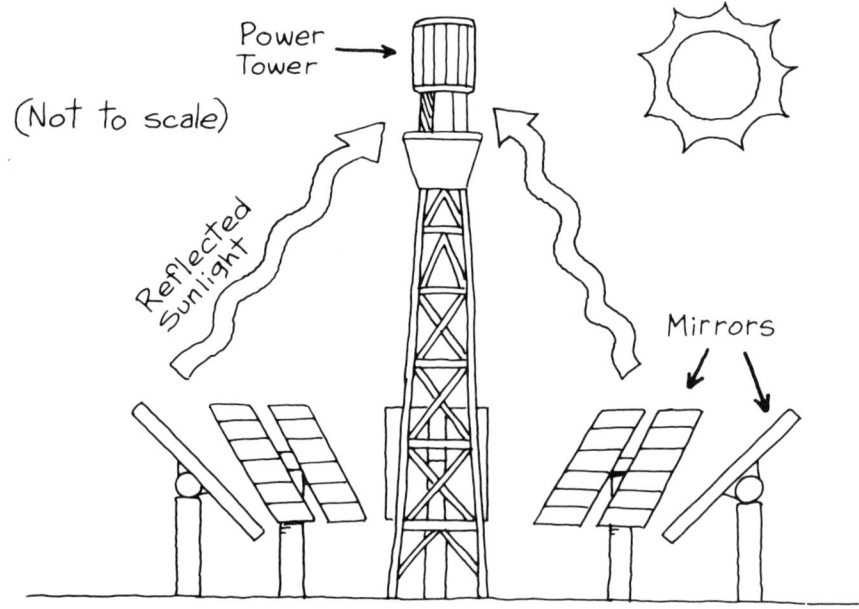

weather is cloudy. The mirrors must be spread out over one square mile of land. It may cost too much to build such a solar power plant to make it worthwhile. The first U.S. solar energy test plant of this kind is in the Southwest. There are no plans yet for widespread use of solar batteries. The cost of solar batteries is much too high to solve our energy needs.

What about wind? Over the centuries windmills pumped water and ground grain into flour. We've already seen that wind can turn generator turbines. The only problem is that the wind doesn't blow fast enough often enough. So engineers are working on ways to store electricity generated when the wind is blowing for use when the air is calm.

Engineers are working on other solutions. Some ideas include using the tides and using the heat that is in the ocean. There are other plans to change coal into clean-burning gas and oil. Still another plan calls for burning garbage as fuel. It is now

clear that there is no one way to make electricity. Today we get electricity with a mix of waterpower, different fuels, and nuclear energy. The future will add more sources of energy to the mix.

Power in the future is a big problem. Thousands of people will help solve it. But that doesn't mean that you can't do something now. You can help cut your electricity bill by turning switches off when you don't need the electricity they control. Doing this also helps save fuel, so that it will last longer. You can tell your family and friends about the energy crisis. Sooner or later people where you live will have to make choices about how their electricity is made. These choices are not simple. Better choices are made when people understand what they will gain and what it will cost.

Join the people who work to solve the energy crisis. Be part of the world that cares about how we get electricity. Make it your business to be in the know about power for the future. And if you do, all I can say is "More power to you!"

How to See a Magnetic Field

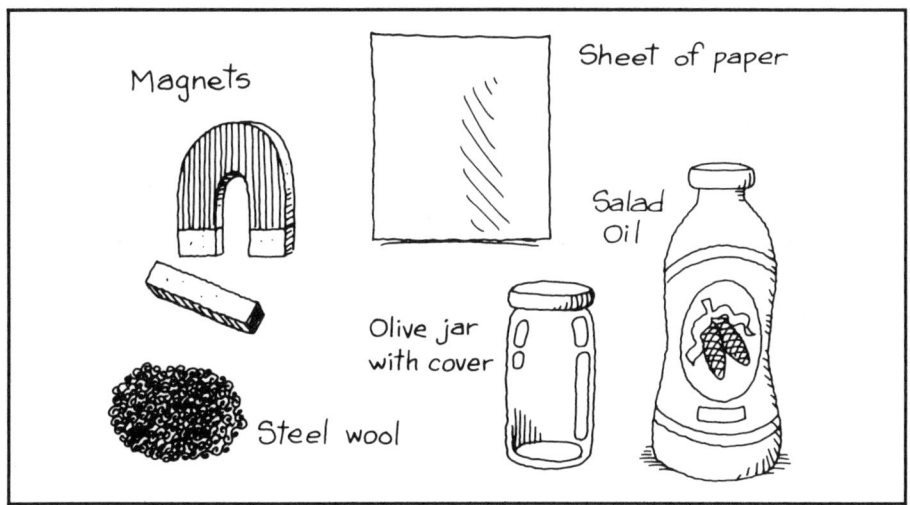

You can't see a magnetic field. But you can detect it. Here are two ways to check out the magnetic fields around magnets. Here's what you need:

- steel wool (from the hardware store, not a soap pad)

- a sheet of paper
- magnets (Check your refrigerator. Some people use magnets to put notices there. You can get magnets at toy stores and at hardware stores.)
- a very small olive jar with a cover
- salad oil

Rest a magnet on a table top. Cover it with a sheet of paper. Hold the steel wool over the paper. Squeeze the steel wool with your hand so that tiny pieces fall on the paper. As bits of steel wool land

on the paper over the magnet, they line up to show the magnetic field. Most of the steel wool particles bunch up at the ends of the magnet. This is where the magnetic field is strongest. The magnetic field extends to the nearby space that has no bits of steel wool in it. All of the pieces that would be there have been pulled over to the end of the magnet.

You can see a magnetic field in space. Fill a small, skinny olive jar with salad oil. Sprinkle about

a tablespoon of steel wool bits into the salad oil. Put the lid on the jar and shake it to spread the steel wool through the oil. Put the end of a magnet on the side of the jar. See how the steel wool lines up with the magnet.

You will see that some magnets collect more steel wool than others. They do this because some magnets are stronger than others. This is one way you can see which are the stronger ones.

LOCUST GROVE ELEM. LIBRARY

2151

DATE DUE

MAR 7						

HIGHSMITH   # 45228